FOSSILS

FOSSILS

CHARTWELL
BOOKS, INC.

Published by Chartwell Books
A Division of Book Sales Inc.
114 Northfield Avenue
Edison, New Jersey 08837
USA

ISBN 0-7858-0970-8

This book is produced by
Quantum Books Ltd
6 Blundell Street
London N7 9BH

Project Manager: Rebecca Kingsley
Project Editor: Judith Millidge
Design/Editorial: David Manson
Andy McColm, Maggie Manson

The material in this publication previously appeared in
Rocks, Shells, Fossils, Minerals and Gems,
On the Trail of the Dinosaurs

QUMSPFS
Set in Futura
Reproduced in Singapore by Eray Scan
Printed in Singapore by Star Standard Industries (Pte) Ltd

Contents

FASCINATING FOSSILS

The way in which a fossil is formed depends on the original chemical composition of the organism, its surrounding matrix and the length of time it has been fossilized. Sometimes chemical changes can either preserve the organism in stone or use the space it used to occupy as a mold, creating an image in rock, millions of years after its death.

How Fossils are Formed

We commonly talk about something old as being a 'fossil', however, the process is more complicated and requires more than simple age.

CONDITIONS FOR FOSSILS

There must be precisely the right set of conditions for the formation of fossils. The dead animal or plant will simply decompose if that combination doesn't exist. A dead animal or plant is attacked by decomposers almost from the time of death, ranging from tiny microbes to larger scavenging vertebrates. Fleshy tissue will disappear first, leaving bones, shells and other hard parts such as teeth. Given time, these vanish as well. Occasionally, however, an organism will settle in a place where oxygen levels are low and the decomposers cannot function.

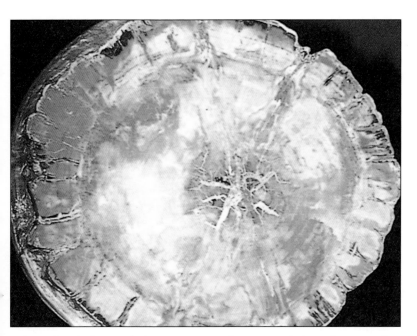

Left. Fossil teeth of C.megalodon suggest it was a 50ft giant Cenozoic shark.

Above. A petrified wood specimen from the Petrified Forest National Park, Arizona.

BURIED ORGANISMS

Organisms may have become buried in a thick layer of fine mud in a lake delta, swallowed by a landslide or shifting sand dune. Or they may become covered by the constant rain of sediment falling on the ocean floor. Here, although the soft parts will soon disappear, any bones, teeth, shell or wood, which are made up of more durable substances such as calcium carbonate or cellulose will take many more years to go.

PETRIFIED FORESTS

In petrified wood specimens the original cellulose has been totally replaced by silica, infiltrating the wood in solution. Sometimes the mineralization only extends to the empty spaces between the wood fibres, while in other cases the wood itself dissolves and is then replaced by minerals. If the wood dissolves, leaving a gap in the harder matrix, and later fills with minerals the result is known as a pseudomorph or natural cast.

Finding Fossils

Fossils are not distributed randomly through the world's rock strata. Certain kinds of rocks yield abundant fossils, while others are utterly barren. Therefore, a basic knowledge of geology will obviously help immensely when you are hunting for fossils.

ROCK VARIETIES

Rocks are found in three basic varieties: igneous, metamorphic and sedimentary. Igneous rocks are formed from crystallized lava and magma. Metamorphic rocks are so named because they have undergone fundamental changes due to pressure and heat. Igneous and metamorphic rocks are both poor hunting grounds for fossils.

Instead, you must search for the sedimentary rocks created when dust, mud, sand or other soil particles collect in layers, then are solidified into rock. The same cloudburst that washes a thick new layer of silt into a lake may also carry down the remains of dead animals and plants, which are buried – these are potential fossils in the making.

Left. This dragonfly larva is a rare insect fossil. Its habitat of muddy pools and rivers is conducive to the fossilization process.

Above. Camerina spp. was one of the Jurassic forams which fell to the sea floor in thick layers helping to form nummulitic limestone .

LIMESTONE FOSSILS

Limestone, such as the white cliffs that are such a feature along the southern coast of England, can provide one of the best hunting grounds for fossils, since it was very often originally created from thick deposits of invertebrate shells many millions of years ago. Nummulitic limestone, for example, comes from ancient deposits of foraminiferan tests, while ancient corals are also recognized as important contributors to many worldwide limestones.

EXPOSED BEDROCK FOSSILS

Digging in rocky soil may turn up an occasional lucky find, but for more consistent results you must look for places where sedimentary bedrock has been exposed, either naturally or, more often, through human activity. These include quarries, road and railway cuttings, mines, tunnels and construction sites as well as natural sites such as outcroppings, fault lines, river banks and coastal and mountain cliffs.

Fossil Field Techniques

Anyone can collect fossils with their bare hands, but the good finds aren't always lying on the surface, weathered free in a handy size. For the more serious hunter, a few pieces of equipment are essential.

CLOTHING

Initially you should start with the right clothing. You will need to wear thick pants, a long-sleeved shirt to help prevent sunburn, a pair of thick work gloves, a brimmed hat and heavy boots. If you are working in a quarry or along a high road, or below cliffs, wear a hard hat and always wear safety goggles when you are splitting or hammering rock.

TOOLS

A strong knapsack, or shoulder bag, should contain a geologist's rock hammer, an assortment of chisels, a good quality hand lens, a small and a medium-size brush (shaving brushes are very highly recommended), a trowel and a sieve; depending on the site and the kinds of fossils you find, a coarse sieve and one with finer mesh may be needed.

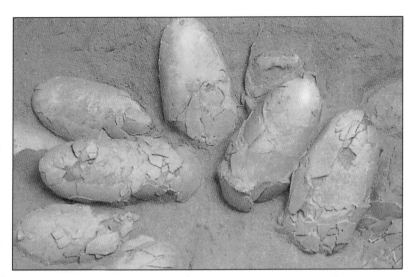

Left. The echinoid, Micraster coranguinum is found in Cretaceous strata in Europe.

Above. Dinosaur eggs of Protoceratops are quite common in the Gobi desert.

FIELD NOTES AND PHOTOS

Your most important pieces of equipment are a notebook and pencil. Good fossil hunters make obsessive notes, recording site location, appearance and layout along with compass bearings and topographical angles. If you are a competent photographer, you can supplement your field notes with clear photographs. But remember, try not to fall into the trap of believing that a photograph can replace field notes. Even a very good photo cannot separate the trivial from the crucial, the way the human eye and mind can.

COLLECTING FOSSILS

Only when preliminary field notes have been taken should you begin removing fossils. Work with the rock, rather than against it, using the hammer as a last resort rather than an initial attack. Many shales will split apart with hand pressure or a prying twist with a chisel or knife, but will shatter irreparably with a hammer blow. Specimens rattling loose around inside your bag will become chipped and scratched, so wrap each fossil in paper and identify temporarily with a number which you can cross-refer to in your field notes.

The Geological Calendar

Palaeontologists use eons, eras, periods and epochs to describe geological time. We are now in the Cenozoic Era, Quaternary Period and Holocene Epoch. Examine the calendar opposite to find our place in time.

HOW LONG IS TIME?

Units of geological time are not of a standard length. The Triassic Period, when the dinosaurs rose to prominence and ammonoids were common in the seas, lasted about 35 million years. Whereas the Jurassic Period, when dinosaurs dominated the earth, lasted about 65 million years.

ERAS, PERIODS AND EPOCHS

To make geological time more manageable, palaeontologists uses eras, periods and epochs. Each unit is shorter than the one before. An epoch being shorter than a period, which in turn is shorter than an era. A complete geological calendar is shown on the facing page.

CENOZOIC ERA

63–65 million years BP to present

Quaternary Period

Holocene Epoch
11,000 years BP to present

Pleistocene Epoch
1.8–2 million years to 11,000 years BP
Climate cools, ice sheets advance.
Homo erectus appears then *Homo sapiens.*

Tertiary Period

Pliocene Epoch
5–7 million years to 1.8–2 million years BP
Grasslands dominate. First hominoids appear.

Miocene Epoch
24–26 million years to 5–7 million years BP
Monkeys and apes appear.

Oligocene Epoch
37 million years to 24–26 million years BP
Grasses, toothed whales appear.

Eocene Epoch
54 million years to 37 million years BP
First horses, elephants, whales, anthropods.

Palaeocene Epoch
65 million years to 54 million years BP
Flightless predatory birds appear.

MESOZOIC ERA

225 million years to 65 million years BP

Cretaceous Period

144 million years to 65 million years BP
Flowering plants dominate. End of dinosaurs.

Jurassic Period

213 million years to 144 million years BP
Dinosaurs dominate. Gondwanaland splits
and continents begin to shift.

Triassic Period

248 million years to 213 million years BP
Dinosaurs rise to prominence. Pangea splits.

PALAEOZOIC ERA

565 million years to 248 million years BP

Permian Period

286 million years to 248 million years BP
Reptiles diversify. Extinction of trilobites.

Carboniferous Period

345 million years to 286 million years BP
Warm, wet climate. Coal-swamp forests.

Devonian Period

395 million years to 345 million years BP
Jawless fishes dominate oceans, with
trilobites, brachiopods and crinoids.

Silurian Period

430 million years to 395 million years BP
Fishes diversify; first primitive land plants.

Ordovician Period

500 million years to 430 million years BP
Jawless fish, molluscs and shellfish appear.

Cambrian Period

565 million years to 500 million years BP
Trilobites and shelled invertebrates appear.

··

THROUGHOUT THE DIRECTORY
The following symbols are used to designate
the Geological Period of each fossil.

QUA	**TERT**	**CRET**
Quaternary	Tertiary	Cretaceous
JUR	**TRI**	**PER**
Jurassic	Triassic	Permian
CAR	**DEV**	**SIL**
Carboniferous	Devonian	Silurian
ORD	**CAM**	
Ordovician	Cambrian	

FASCINATING FOSSILS

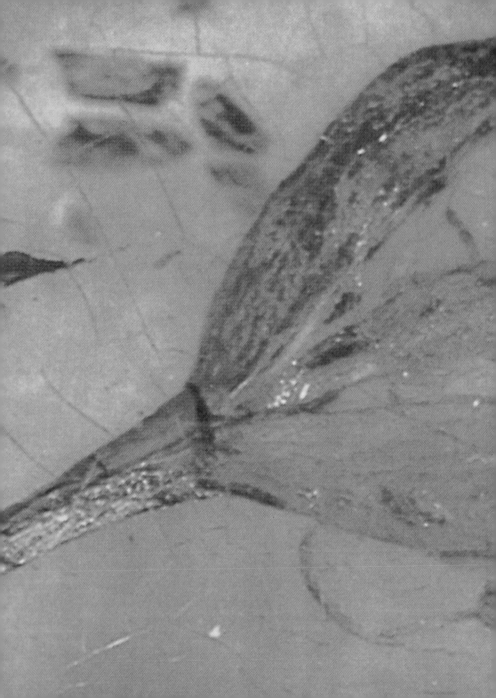

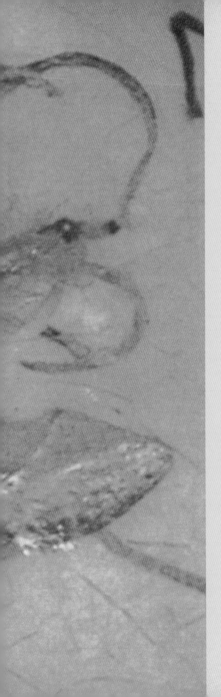

FOSSIL SPECIMENS

Key to symbols

The following icons are used throughout the directory to indicate the organism group of each fossil.

Protozoa.

Plants.

Invertebrates.

Vertebrates.

CAMERINA SPP

Important out of all proportion to their size, these often microscopic protozoans were and are plentiful in the oceans. Their limey tests, falling to the sea floor in thick layers, are a major contributor to the formation of limestone, and are so common that they serve as valuable indicator fossils for palaeontologists, recording past climates.

Organism group Foraminiferans.
Period Jurassic–Miocene.
Location Europe, South Pacific, Gulf of Mexico.

HYMENAEA SPP

First appearing in the fossil record roughly 100 million years BP, flowers represented a major advance in plant reproduction. By attracting insects, the plants enjoyed a much higher rate of pollination success. The earliest angiosperms, or flowering plants, were the magnolias. Shown here is a flower from a species of Hymenaea, a member of the legume family.

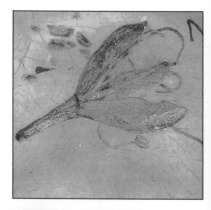

Organism group Angiosperms.
Period Cretaceous.
Location Worldwide.

CALAMITES

Horsetails were an important part of the Carboniferous forests; this type, one of the more common fossils from the period, reached heights of 60ft. The fossils represent stems and were longitudinally ridged. Horsetails remain common wetland plants in many parts of the world, but surviving species are small, usually no more than waist-high.

Organism group Gymnosperms.
Period Upper Carboniferous.
Location Worldwide.

ANNULARIA

The whorled, slightly spatulate leaves of Annularia are common in Pennsylvanian deposits. In the specimen shown, the leaf clusters have been dislodged and have fossilized individually, but they are frequently found still attached to the upper surface of the twig.

Organism group Gymnosperms.
Period Upper Carboniferous.
Location Worldwide.

STIGMARIA FICOIDES

Stigmaria are the fossilized roots of tree clubmosses, *Lepidodendron*. While most surviving clubmosses are small, this genus attained heights of more than 150ft, with a heavily scaled trunk and wide-spreading root system. The genus is extinct but numbered more than 100 species at its peak. Stigmarid roots are usually found as casts, in sediment layers immediately beneath coal seams.

Organism group Gymnosperms.
Period Carboniferous.
Location Europe.

  CAR

PECOPTERIS SPP

This was a true fern, or pteridophyte, although there is a suspicion that some members of the genus had developed into seed-ferns. True ferns represented several 'firsts' in evolutionary history, including the first complex leaves and branched stems. Ferns today are much smaller, inhabiting the shade of the more advanced angiosperm trees.

Organism group Gymnosperms.
Period Carboniferous.
Location Europe, North America, North Africa, Asia.

   CAR

ALETHOPTERIS SPP

The coal-swamp forests of the Carboniferous covered most of what is now the Northern Hemisphere. *Alethopteris* is a seed-fern (pteridosperm) relatively thin, tapered leaflets that are straighter than many of its close relatives. Seed-ferns were a big step forward in plant evolution. The plant embryo was enclosed and protected, germinating when conditions were most favorable.

Organism group Gymnosperms.
Period Carboniferous.
Location USA, Europe, Asia.

 JUR

GLOSSOPTERIS

This is a critically important fossil which contributed to the make-up of the prehistoric supercontinent known as Gondwanaland. It has been found in Antarctica, which had a subtropical climate at the time. Most fossils are of single leaves, lanceolate or spatulate. This was a seed-fern that grew into tree form. This specimen is from New South Wales.

Organism group Gymnosperms.
Period Triassic.
Location Antarctica, New South Wales (Australia).

 TRI

PETRIFIED WOOD

Petrified wood is an unusual fossil with a wide distribution; there are significant sites in North America and Europe. This specimen comes from the Petrified Forest National Park in Arizona, where primitive protopines, *Araucarioxylon*, were buried in saturated sediment and replaced with agate and jasper.

Organism group Gymnosperms.
Period Triassic.
Location Worldwide.

 TRI

AMBER

Many trees, particularly conifers, secrete a resin when injured. The resin seals the wound against insects, disease and desiccation, and under some conditions it may harden to form a globule of amber. Insects and arachnids often become mired in tree resin, and those trapped in amber display an astounding degree of preservation.

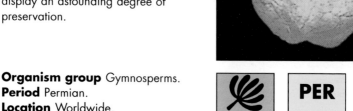

Organism group Gymnosperms.
Period Permian.
Location Worldwide.

PER

SIPHONIA SPP

Flower-shaped, the stalked sponges of the genus *Siphonia* had siliceous spicules, a large central cavity in the bud-like body, and a network of tiny openings that connected to a primitive vascular system. *Siphonia* were 'rooted' at the end of the stalk to the sea bed, where the ocean currents brought them food particles.

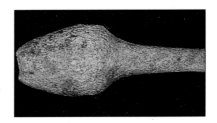

Organism group Invertebrates.
Phylum Porifera (sponges).
Period Cretaceous.
Location Europe.

 CRET

VENTRICULITES SPP

Species of the genus *Ventriculites* show a distinctive, vase-shaped body with a fused spicule construction; the body walls are thin and have large pores. This genus belongs to the group of sponges known as lychniskids, in which diagonal braces buttress the form, creating a pattern similar to a Greek lantern, on which the group name is based.

Organism group Invertebrates.
Phylum Porifera (sponges).
Period Cretaceous.
Location Europe.

 CRET

RAPHIDONEMA SPP

Unlike the previous sponges, the sponges of the genus *Raphidonema* have calcareous spicules with three limbs, and are fused into a rigid skeleton. The overall appearance of the sponge is that of an open, widely flared vase, often heavily folded and covered with large pores and lumps. The species shown is *R. parcatum.*

Organism group Invertebrates.
Phylum Porifera (sponges).
Period Cretaceous.
Location Europe.

 CRET

HALYSITES SPP

Often known as chain coral, *Halysites* is one of the tabulate corals; the specimen shown, *H. cantenularius,* has been cut in cross-section, revealing the individual corallites with their tabulae, the horizontal sections that give this group its name. In life a *Halysites* colony would have been a series of long, thin, upright tubes, joined along the edges.

Organism group Invertebrates.
Phylum Cnidaria (corals, jellyfish, hydrozoans).
Period Silurian.
Location Worldwide.

 SIL

FAVOSITES SPP

One of the most common fossil corals of
the Middle Palaeozoic, *Favosites* is
a tabulate coral with irregularly
five-sided corallites in cross-
section. The walls of the
corallite are thin and pocked
with mural pores. The corallum
may be large or irregularly
shaped and known as honey-
comb coral because of the shape.

Organism group Invertebrates.
Phylum Cnidaria (corals, jellyfish,
hydrozoans).
Period Silurian–Devonian.
Location Worldwide.

 SIL

ZAPHRENTIS SPP

One of the rugose, or horn, corals,
Zaphrentis was a solitary type that
secreted a cup-shaped corallite.
The pointed end of the coral
would have originally been
anchored in mud, with the open
end of the cup facing upwards
so the coral animal could filter
food particles from the water.

Organism group Invertebrates.
Phylum Cnidaria (corals, jellyfish,
hyrozoans).
Period Devonian-Lower Carboniferous.
Location Europe, North America.

 DEV

CNIDARIA

ACERVULARIA SPP

One of the colonial rugose corals, *Acervulvaria* formed large, usually four-sided corallites with pronounced finely toothed septae. The species shown here, *A. ananas*, known as the pineapple coral, is common in the Wenlockian sediments of England. The rugose corals had six major septae.

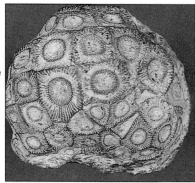

Organism group Invertebrates.
Phylum Cnidaria (corals, jellyfish, hydrozoans).
Period Silurian–Devonian.
Location Europe.

 SIL

LITHOSTROTION SPP

The corallites may be pentagonal, round or four-sided, and have cone-shaped tabulae and a conical lump in the centre of the calice, ridged by septae. This genus was a colonial rugose coral, with colonies forming a significant portion of the fossil reefs of the Carboniferous.

Organism group Invertebrates.
Phylum Cnidaria (corals, jellyfish, hydrozoans).
Period Devonian–Lower Carboniferous.
Location Northern Hemisphere, Africa, Australia.

 DEV

PALAEOSMILIA SPP

One of the rugose corals, *Palaeosmilia* is found in both haploid (solitary) and compound (colonial) forms. The hallmark of this genus is the highly packed septae; in *Palaeosmilia* the septae are radially symmetrical, rather than bilaterally as in most other rugose corals.

Organism group Invertebrates.
Phylum Cnidaria (corals, jellyfish, hydrozoans).
Period Carboniferous.
Location Worldwide.

 CAR

SCLERACTINID CORALS

The scleractinids include the sea anemones and the stony corals, the latter being far more important to the fossil record than the soft, rarely fossilized anemones. Scleractinids – or hexacorals – may be solitary or colonial, and include all the world's living hard corals. Shown is an unidentified species of scleractinid from the Jurassic.

Organism group Invertebrates.
Phylum Cnidaria (Corals, jellyfish, hydrozoans).
Period Triassic–Recent.
Location Worldwide.

 TRI

FENESTELLA SPP

This common lace bryozoan formed colonies that fanned out, with thin cross-bars linking the branches. Each branch possessed twin rows of zooids, on one side, so that a pseudomorph of the wrong side of a *Fenestalla* colony may not show the zooids. The species shown is *F. plebeia* from the Lower Carboniferous.

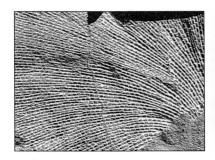

Organism group Invertebrates.
Phylum Bryozoa (bryozoan).
Period Ordovician-Permian.
Location Europe.

 ORD

OBOVTHYRIS SPP

Bulbous in shape, *Obovothyris* is a rib-less brachiopod with a small pedicle foramen and a strongly curved umbo, the 'point' or beak of the pedicle valve. The species illustrated is *O. magnobovata.*

Organism group Invertebrates.
Phylum Brachiopoda (brachiopods).
Period Jurassic.
Location Europe.

 JUR

SPHAEROIDOTHYRIS SPP

Similar to *Obovothyris*, *Spaeroidothyris* is another bulbous Jurassic brachiopod with poorly defined growth lines on the valve. The species shown is the type, *S. sphaeroidothyris*.

Organism group Invertebrates.
Phylum Brachiopoda (brachiopods).
Period Jurassic.
Location Worldwide.

 JUR

RHYNCHONELLA SPP

The type genus for the rhynchonellids, *Rhynchonella* has a well-developed pedicle foramen wedged between a pair of deltidial plates on the umbo. Most species have strong, radial ribbing, with a simplistic internal support for the lophophore. Surviving rhynchonellids differ little from those fossilized more than 500 million years BP. This specimen is *R. capex* from the Silurian

Organism group Invertebrates.
Phylum Brachiopoda (brachiopods).
Period Ordovician–Recent.
Location Europe.

 ORD

LOBOTHYRIS SPP

Lobothyris is one of the terebratulid brachiopods, which have strongly biconvex shells and pedicle foramen. *Lobothryis* mastered an environment that other brachiopods could not tolerate, and is found in great numbers. Lack of competition meant that they dominated their particular niche, which today has been transformed into ironstone deposits. The species shown is *L. punctata*.

Organism group Invertebrates.
Phylum Brachiopoda (brachiopods).
Period Jurassic.
Location Worldwide.

 JUR

RUGITELA SPP

Rugitela has distinct growth lines on its valves, which reach a maximum size of 1¹/₂in in large specimens. The foramen is small and the umbo long and sharply hooked, curving back on the brachial valve.

Organism group Invertebrates.
Phylum Brachiopoda (brachiopods).
Period Jurassic.
Location Worldwide.

 JUR

GIBBITHYRIS SPP

Globular and smooth, this small brachiopod has biconvex valves and a small pedicle foramen. Growth lines are closely spaced on *Gibbithyris,* a common fossil from England. It belongs to a group of brachiopods known as terebratulids, which feature strong umbones and hinge lines less than the width of the valves at their widest point. Most of the surviving brachiopods are terebratulids.

Organism group Invertebrates.
Phylum Brachiopoda (brachiopods).
Period Cretaceous.
Location England.

  CRET

PLECTOTHYRIS SPP

A small brachiopod, *Plectothyris* has biconvex valves and a well-developed umbo pierced by the pedicle foramen. The valves are ribbed. The tip of the umbo has been cracked off in this specimen, revealing plain rock – for this, like many fossils, is a natural cast, retaining only the external shape of the animal that made the mold before dissolving away.

Organism group Invertebrates.
Phylum Brachiopoda (brachiopods).
Period Jurassic.
Location Worldwide.

  JUR

B
R
A
C
H
I
O
P
O
D
A

TURRITELLA SPP

A long, tapered spire with slightly flattened whorls distinguishes *Turritella,* a common fossil in strata laid down in shallow oceans. Depending on the species, the shell is 2in or less in length with ribbed or keeled whorls, and the aperture varying from round to square. Also known as turret shells.

Organism group Invertebrates.
Phylum Mollusca (molluscs).
Class Gastropoda (gastropods).
Period Cretaceous-Recent.
Location Worldwide.

  CRET

VIVIPARUS SPP

Viviparus snails had smooth, rounded whorls and moderate spires, with large, circular apertures. The photograph shows specimens of the species *V. sussexiensis* from Europe. A surviving member of the species *V. georgianus,* the Georgia apple snail, is a common freshwater species in the southern USA.

Organism group Invertebrates.
Phylum Mollusca (molluscs).
Class Gastropoda (gastropods).
Period Cretaceous–Recent.
Location Europe.

  CRET

NEPTUNEA SPP

Still an abundant genus in the world's oceans, *Neptunea* whelks first appear in the fossil record in the Cretaceous; the photographs shows *N. contraria* of the Pleistocene. The shells have large, smooth body whorls and moderately tapered spires; the aperture is oval, forming a teardrop where it meets the siphonal canal.

Organism group Invertebrates.
Phylum Mollusca (molluscs).
Class Gastropoda (gastropods).
Period Cretaceous–Recent.
Location Worldwide.

 CRET

SCAPHELLA SPP

Living *Scaphella*, known as volutes, are colorfully marked shells popular with collectors. Fossils of this shell usually have nodes on the whorls, and four folds on the columella. The species shown is *S. lamberti* from the Pleistocene. In *Scaphella*, the first secretions are horny and temporary, dropping off after the calcareous shell is secreted, leaving a small point as evidence of their presence.

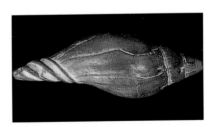

Organism group Invertebrates.
Phylum Mollusca (molluscs).
Class Gastropoda (gastropods).
Period Cretaceous–Recent.
Location Worldwide.

 CRET

ATHLETA SPP

A robust fossil whelk, *Athleta* has weak to strong nodes which may form short spines, and a long, slightly curved siphonal canal. The body whorl is ribbed. This genus includes a wide variety of living species, while fossil forms were especially important during the Cretaceous and Tertiary. The species shown is *A. luctator* from the Eocene.

Organism group Invertebrates.
Phylum Mollusca (molluscs).
Class Gastropoda (gastropods).
Period Cretaceous.
Location Europe.

CRET

CLAVILITHES SPP

This genus has a long spire, shouldered whorls and faint longitudinal ridging on the whorls. The long siphonal canal is missing from this specimen of *C. macrospira* from the Eocene. The protoconch on *Clavilithes* is a form known as mammillated, since it takes the shape of two squat, rounded whorls, which are unornamented.

Organism group Invertebrates.
Phylum Mollusca (molluscs).
Class Gastropoda (gastropods).
Period Cenozoic.
Location North America, Europe, Asia.

QUA

CONUS SPP

So widespread that they serve as index fossils for the Cenozoic, *Conus* shells have flat to short spires with a long narrow aperture slit at each end. The whorls are smooth except for growth lines. They remain a diverse genera of gastropods, most are found in shallow inshore waters. The venom of some Indo-Pacific species, like the textile cone, has caused human deaths.

Organism group Invertebrates.
Phylum Mollusca (molluscs).
Class Gastropoda (gastropods).
Period Cretaceous–Recent.
Location Worldwide.

 CRET

APTYXIELLA SPP

One of the most graceful of Jurassic fossils, *Aptyxiella* has a long spire, which in some species becomes tapered, with a tiny aperture and slight indentations on the insides of the whorls. The specimen shown, of *A. porlandica*, are natural casts, formed when sediment entered the shell; the shells later dissolved, leaving a replica of their interiors.

Organism group Invertebrates.
Phylum Mollusca (molluscs).
Class Gastropoda (gastropods).
Period Jurassic–Cretaceous.
Location Europe.

 JUR

PLEUROTOMARIA SPP

The shell of *Pleurotomaria* forms a broad cone, with slight knobbing on the shoulders and growth lines on the whorls that give the appearance of ridging; the aperture is rounded, with a pronounced slit in the outer lip. This is a common gastropod fossil in many Upper Mesozoic deposits worldwide. The specimen shown is *P. bitorquata*.

Organism group Invertebrates.
Phylum Mollusca (molluscs).
Class Gastropoda (gastropods).
Period Jurassic–Cretaceous.
Location Worldwide.

 JUR

CERITHIUM SPP

A small shell, *Cerithium* has a long spire, with a sharp apical angle, aperture lips that may flare, and a short, curved siphonal canal; the whorls may be knobbed or ridged with growth lines, or flat and unornamented. Modern species, known as ceriths, are highly ornamented and colorful, and inhabit ocean shallows and reefs. Shown is *C. duplex* from Europe.

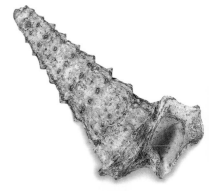

Organism group Invertebrates.
Phylum Mollusca (molluscs).
Class Gastropoda (gastropods).
Period Cretaceous–Recent.
Location Worldwide.

 CRET

ANTHRACOSIA AND CARBONICOLA SPP

These two small species were freshwater inhabitants, living in the coal-swamp forests and marshes of the Carboniferous, and among the first bivalves to make the change from saltwater to fresh. The *Carbonicola* shell is subtriangular, with a curved hinge line and minor growth-line ridging. *Anthracosia* belongs to the same family and is similar. Shown are *A. atra* (left) and *C. communis* (right).

Organism group Invertebrates.
Phylum Mollusca (molluscs).
Class Bivalvia (bivalves).
Period Carboniferous.
Location Europe.

CARDIUM SPP

One of the bivalves known as cockles, *Cardium* has valves that appear heart-shaped in profile, with pronounced ribbing on the external surfaces and distinct muscle scars on the interior. Illustrated is *C. parkinsoni* from the Pleistocene. *Cardium* species are heterodonts, having lateral teeth on the hinge, as well as cardinal teeth.

Organism group Invertebrates.
Phylum Mollusca (molluscs).
Class Bivalvia (bivalves).
Period Triassic–Recent.
Location Worldwide.

MYPHORELLA SPP

The distinguishing feature of this Upper Mesozoic bivalve is the concentric rows of knobs that cover the triangular to subtriangular valves, with a sharp beak at the dorsal margin. The species shown above is *M. budlestoni*; these bivalves are sometimes assigned to the genus *Trigonia* and were burrowing shellfish.

Organism group Invertebrates.
Phylum Mollusca (molluscs).
Class Bivalvia (bivalves).
Period Jurassic–Cretaceous.
Location Worldwide.

 JUR

PLEUROMYA SPP

As *Pleuromya* species are desmodont bivalves, with no true hinge teeth, it is believed that they were burrowers, since a buried bivalve has little need for an efficient open-and-close mechanism in the shell. *Pleuromyra* species are equivalve bivalves, oval to trapezoidal in shape. There may be concentric ribbing, which is more pronounced in some specimens than others.

Organism group Invertebrates.
Phylum Mollusca (molluscs).
Class Bivalvia (bivalves).
Period Triassic–Cretaceous.
Location Worldwide.

 TRI

GLYCYMERIS SPP

A common genus in the world's
oceans, *Glycymeris* is also abundant
in the fossil record from the Cretaceous
on. Most species are circular to slightly
oval, with large umbones; external
ornamentation comprises costae,
varying in development, and growth
lines. Shown is *G. delta* from the
Eocene.

Organism group Invertebrates.
Phylum Mollusca (molluscs).
Class Bivalvia (bivalves).
Period Cretaceous–Recent.
Location Worldwide.

  CRET

ARCTICA SPP

Clams of the genus *Arctica* showed a
preference for cold waters, making them
valuable as environmental indicators. The
valves are typical of cold-water clams –
oval to circular, with curved umbones
and no external ornamentation beyond the
growth lines. Shown here is a Pliocene
specimen of *A. islandica*, known as the
black clam, that often appears as a
beach shell on North American shores.

Organism group Invertebrates.
Phylum Mollusca (molluscs).
Class Bivalvia (bivalves).
Period Cretaceous–Recent.
Location North America, Europe.

  CRET

NEITHEA SPP

Found worldwide in Cretaceous deposits, *Neithea* is a lovely inequivalve genus, with its right valve far more convex than the left, which is flattened. The ribbing pattern is distinctive and attractive, with four to six large, roughly equal ribs, creating a scalloped pattern along the anterior margin. The species shown is *N. sexcosta*.

Organism group Invertebrates.
Phylum Mollusca (molluscs).
Class Bivalvia (bivalves).
Period Cretaceous.
Location Worldwide.

 CRET

VENERICARDIA SPP

Large cockles with heavy, triangular valves, low, flat costae and concentric growth lines, *Venericardia* species have two large, curved, hinge teeth. There are a number of living, cold-water species, found from the lower intertidal zone to several hundred feet of water. The fossil specimens shown are *V. planicosta* from the Eocene.

Organism group Invertebrates.
Phylum Mollusca (molluscs).
Class Bivalvia (bivalves).
Period Cretaceous/Palaeocene–Eocene.
Location Europe, Africa, North America.

  JUR

PLICATULA SPP

This species are typical isodonts, bivalves with two identical teeth and two sockets in each valve, placed symmetrically. Shown is a Jurassic species, *P. spinosa*. *Plicatula* species cement themselves in place, securing a foothold. The wave action requires a bivalve to glue itself in place or else be swept away.

Organism group Invertebrates.
Phylum Mollusca (molluscs).
Class Bivalvia (bivalves).
Period Jurassic–Recent.
Location Worldwide.

 JUR

LOPHA SPP

A strongly curved oyster, *Lopha* valves are thickened, with extremely pronounced ribs that form heavy serrations. The thickening is believed to be a defence against carnivorous gastropods, and is still used by living *Lopha* species. The specimen shown is *L.carinata*, seen in profile.

Organism group Invertebrates.
Phylum Mollusca (molluscs).
Class Bivalvia (bivalves).
Period Triassic–Recent.
Location Worldwide.

 TRI

GERVILLELLA SPP

Elongated and lanceolate, the valves of *Gervillella* are pointed anteriorly, forming a dagger-like shape, but they are unornamented except for concentric growth lines. A medium or large inequivalve, *Gervillella* is found worldwide in appropriate deposits.

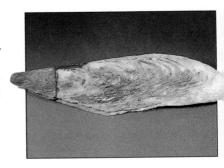

Organism group Invertebrates.
Phylum Mollusca (molluscs).
Class Bivalvia (bivalves).
Period Triassic–Cretaceous.
Location Worldwide.

 TRI

EXOGYRA SPP

Exogyra species are inequivalved fossils of the group loosely known as coiled oysters. The left valve is enlarged with a curled umbo, while the right valve is reduced and flat. Members of this genus were sedentary, remaining cemented to the ocean floor. The specimen shown is *E. latissima*.

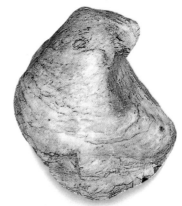

Organism group Invertebrates.
Phylum Mollusca (molluscs).
Class Bivalvia (bivalves).
Period Cretaceous.
Location Worldwide.

 CRET

VENUS SPP

A large and heavy-shelled clam, *Venus* is oval to circular, with growth lines that form raised, sharp-edged concentric ridges on the circular to oval valves. The species shown is *V. casina* from the Pliocene deposits. The quahog, of New England clam chowder fame, was once classified with the *Venus* clams, but has been reassigned to the genus *Mercenaria*.

Organism group Invertebrates.
Phylum Mollusca (molluscs).
Class Bivalvia (bivalves).
Period Oligocene–Recent.
Location Worldwide.

 TERT

SPONDYLUS SPP

An extremely widespread fossil genus, *Spondylus* is an inequivalve with irregular costae and, depending upon the species, irregular spines or knobs growing from the costae. The right valve is cemented to the substrata. One of the most dramatic members of this genus is the Atlantic thorny oyster (*S. americanus*), which develops long, delicate spines.

Organism group Invertebrates.
Phylum Mollusca (molluscs).
Class Bivalvia (bivalves).
Period Jurassic–Recent.
Location Worldwide.

 JUR

ORTHOCERAS SPP

The most famous shell of the fossil
cephlopods, the nautilus and ammonites,
have curved shells, but the Palaeozoic
produced many species with straight or
only slightly curved shells. This specimen,
from Morocco, shows an *Orthoceras*
shell in cross-section. The septae dividing
the body chambers are visible and the
tube-like siphuncle running the length of
the shell can be seen clearly.

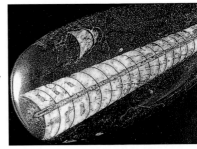

Organism group Invertebrates.
Phylum Mollusca (molluscs).
Class Cephalopoda (cephalopds).
Period Ordovician–Triassic.
Location Europe, North Africa.

 ORD

NAUTILUS SPP

More than 180 million years old, this
fossil *Nautilus* echoes the sole surviving
species of its genus, the chambered
nautilus. In *Nautilus,* the shell coils up
and behind the body chamber like a
ram's horn; the live animal, which was
tentacled like a squid, protruded from the
aperture, and was able to move by jet
propulsion, squirting water through
gas-filled chambers inside the shell.

Organism group Invertebrates.
Phylum Mollusca (molluscs).
Class Cephalopda (cephalopods).
Period Oligocene–Recent.
Location Europe, Australia, East Indies.

 TERT

DACTYLIOCERAS SPP

A famous, commonly collected ammoniod of Eurasia and Latin America, *Dactylioceras* possesses evolute whorls and simple, almost straight ribs. There is no keel, and the sutures are lobed. Shown are three specimens of the species *D. commune,* an important fossil in European strata, in which the ribs divide over the ventral surface of the whorls.

Organism group Invertebrates.
Phylum Mollusca (molluscs).
Class Cephalopoda (cephalopods).
Period Jurassic.
Location Worldwide.

 JUR

HILDOCERAS SPP

An evolute ammonoid with flattened whorls, *Hildoceras* has a shallow groove that runs the length of each side of the whorls as well as along the keel; the ribs are curved. The specimen illustrated is *H. bifrons.*

Organism group Invertebrates.
Phylum Mollusca (molluscs).
Class Cephalopoda (cephalopods).
Period Jurassic.
Location Worldwide.

 JUR

AMALTHEUS SPP

Amaltheus belongs to a worldwide group of ammonoids known as ammonitids, a suborder with simplistic sutures, heavily ornamented shells and – in some cases – gigantic sizes, with diameters of 3ft. This species is much smaller, with an average size of 3in. The whorl coiling is involuted, a keel is present, and the ribs form gentle S-shapes. Shown is the species *A. margarinatus*.

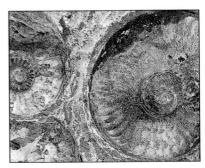

Organism group Invertebrates.
Phylum Mollusca (molluscs).
Class Cephalopoda (cephalopods).
Period Jurassic.
Location Worldwide.

 JUR

ARNIOCERAS SPP

A common Jurassic ammonoid, *Arnioceras* has strong, straight ribs that curve forward on the venter, stopping at the keel. The coiling is evolute, and the saddles and lobes of the sutures are themselves convoluted into smaller subdivisions. The specimen shown is *A. semicostatum* from the Lower Jurassic, the variety most commonly found.

Organism group Invertebrates.
Phylum Mollusca (molluscs).
Class Cephalopoda (cephalopods).
Period Jurassic.
Location North and South America, Europe, Africa, Asia.

 JUR

NIOBELLA SPP

A mid-sized trilobite, with lengths of up to 4in. *Niobella* fossils show a well-developed cephalon, large pygidium and eyes. There are eight thoracic segments. Like all trilobites, it was benthic – that is, spending virtually all its time on the ocean floor, judging from the shape of its appendages, and its digestive system. The species shown is *N. homfrayi*.

Organism group Invertebrates.
Phylum Arthropoda (arthropods).
Class Trilobita (trilobites).
Period Ordovician.
Location Worldwide.

OGYGIOCARELLA SPP

Trilobites of the genus *Ogygiocarella* are common in European and Latin American deposits from the Ordivician. The genus is macropygous – that is the pygidium is almost as large as the thorax; in addition, the eye is short and curved, the gladbella bulges out anteriorly, and the facial suture cuts the rear border of the cephalon. The species shown is *O. debuchi*.

Organism group Invertebrates.
Phylum Arthropoda (arthropods).
Class Trilobita (trilobites).
Period Ordovician.
Location Europe, Latin America.

ELRATHIA SPP

A North American trilobite, *Elrathia* is an important member of the Cambrian fauna, with the species *E. kingii* (shown here) being especially abundant and widespread. They belonged to a group of trilobites called opisthoparians, named for their distinct head sutures. The largest of the trilobites, the opisthoparians could reach more than 2ft in length.

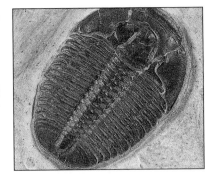

Organism group Invertebrates.
Phylum Arthropoda (arthropods).
Class Trilobita (trilobites).
Period Cambrian.
Location North America.

CAM

PHACOPS SPP

These were the dominant genera of trilobites from Middle Palaeozoic rocks in North America. *Phacops rana,* shown here, is typical; the glabella is large, with at least two furrows. The most commonly found fossils are those of shed pieces of exoskeltons. Fossil trilobite eyes retain the faceted surface that in life was a compound lens.

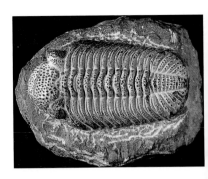

Organism group Invertebrates.
Phylum Arthropoda (arthropods).
Class Trilobita (trilobites).
Period Silurian–Devonian.
Location Worldwide.

SIL

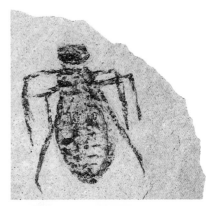

LIBELLULA DORIS

Insects are rare as fossils, partly because of their thin, easily destroyed exoskeletons. This dragonfly larva is one of the unusual exceptions, perhaps because its habitat – ponds or muddy bottomed rivers – is more conducive to fossilization. The dragonfly family *Libellulidae* remains widespread today.

Organism group Invertebrates.
Phylum Arthropoda (arthropods).
Class Insecta (insects).
Period Miocene.
Location Worldwide.

 **TERT**

INSECT TRAPPED IN AMBER

Almost 40 million years ago, this insect became trapped in tree resin on the island of Dominica. Eventually entombed, it quickly died of suffocation, and the complete lack of oxygen that killed it also prevented bacteria from decomposing it. Eventually the resin fossilized into amber, preserving the insect. Although appearing intact, it is really a hollow mold in the resin, with a carbonized film at the edges.

Organism group Invertebrates.
Phylum Arthropoda (arthropods).
Class Insecta (insects).
Period Tertiary.
Location Dominica.

 TERT

CYATHOCRINITES SPP

A common Palaeozoic crinoid, *Cyathocrinites* has a globular calyx and branching arms composed of a single row of plates, as can be seen in this specimen of *C. arthritcus,* from the Silurian. Stalked crinoids like this were dominant in the ancient seas of the Palaeozoic, but they have been replaced today, by stalkless, mobile genera. Stalked crinoids are only found on deep reefs in tropical regions.

Organism group Invertebrates.
Phylum Echinodermata (echinoderms).
Class Crinoidea (crinoids).
Period Silurian–Carboniferous.
Location North America.

 SIL

BOTRYOCRINUS SPP

Crinoids were far more successful than their close relatives, the cystoids and blastoids, outnumbering them nearly five to one in terms of fossil genera, with a total of more than 750. Only the crinoids survived the mass extinctions at the end of the Permian. This species of *Botryocrinus* is from the Wenlock limestones.

Organism group Invertebrates.
Phylum Echinodermata (echinoderms).
Class Crinoidea (crinoids).
Period Silurian.
Location Europe.

 SIL

HEMICIDARIS SPP

The widespread fossil sea urchin is usually preserved as the test alone. The test is knobbed with tubercles like a floating mine, and these become larger as they approach the underside. The test is a flattened sphere. The ambulacral grooves are ornamented with rows of tiny tubercles.

Organism group Invertebrates.
Phylum Echinodermata (echinoderms).
Class Echinoidea (sea urchins, sand dollars and allies).
Period Jurassic–Cretaceous.
Location Worldwide.

 JUR

PYGURUS SPP

The star-shaped ambulacral grooves make *Pygurus* instantly recognizable as a 'sand dollar', the common name for the tests of flattened, short-spines sea urchins. The edges of the ambulacral grooves are studded with pores, through which tube feet protruded. In life, *Pygurus* would probably have been buried slightly in sand, allowing wave action to bring food to it.

Organism group Invertebrates.
Phylum Echinodermata (echinoderms).
Class Echinoidea (sea urchins, sand dollars and allies).
Period Cretaceous–Eocene.
Location Worldwide.

 CRET

MICRASTER SPP

Echinoids can be regular or irregular, this specimen, *Micraster,* is irregular; the test is heart-shaped, with short ambulacral plates. A groove leads from the centre of the test to the oral opening on the side, with a low ridge running in the direction, This genus is used to identify cretaceous strata in Europe. The species shown here is M. *coranguinum.*

Organism group Invertebrates.
Phylum Echinodermata (echinoderms).
Class Echinoidea (sea urchins, sand dollars and allies).
Period Cretaceous–Palaeocene.
Location Worldwide.

CRET

PLEGIOCIDARIS SPP

Each genus of echinoid has a distinctive spine shape that helps in identification. They are usually found with other inhabitants of ancient coal reefs, and it serves as an environmental indicator fossil, especially with those of warm, shallow waters.

Organism group Invertebrates.
Phylum Echinodermata (echinoderms).
Class Echinoidea (sea urchins, sand dollars and allies).
Period Jurassic.
Location Worldwide.

JUR

PHYLLOGRAPTUS SPP

The *Phyllograptus* rhabdosome has four jointed stipes, creating an X-shape when sectioned. The thecae of this genus are tubular and simplistic. Like most graptolites, the stipes are thinner than a pencil lead.

Organism group Invertebrates.
Phylum Hemichordata.
Class Graptolthina (graptolites).
Period Ordovician.
Location Worldwide.

MONOGRAPTUS SPP

A close-up of a *Monograptus* stipe shows the tightly packed, uniserial theca. This is a highly variable genus, with stipe and thecae shape ranging from straight to gracefully curved. *Monograptus* is one of the most commonly found of the graptolites.

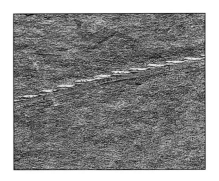

Organism group Invertebrates.
Phylum Hemichordata (Hemichordates).
Class Graptolthina (Graptolites).
Period Silurian–Devonian.
Location Worldwide.

ODONTASPIS SPP

The sand sharks of the genus *Odontaspis* possess long, curved teeth that are often found along ocean beaches, particularly those in Florida. The quantity of fossilized teeth can give a false idea of the numbers of sharks in Cenozoic seas. Sharks loose their teeth constantly throughout their lives, so one shark can produce hundreds in its lifetime.

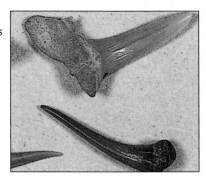

Organism group Vertebrates.
Class Chondrichthyes (Cartilaginous fish).
Period Jurassic–Recent.
Location Worldwide.

 JUR

CARCHARODON SPP

The size, shape and serrated edges of this tooth, show it belonged to *C. megalodon*, a Cenozoic relative of the modern great white shark. Comparing the size of teeth from *C. megalodon* to those of living great whites, it is believed that the extinct species grew to at least 50ft long.

Organism group Vertebrates.
Class Chondrichthyes (Cartilaginous fish).
Period Cretaceous–Recent.
Location Worldwide.

  CRET

ACANTHONEMUS SPP

This fossil fish, a rarity in Tertiary strata, appears to be related to the pompanos and crevalles, and is classified with the Perciformes. The species shown is *A. subaureus.* The genus is found from the Eocene to the Oligocene in European strata.

Organism group Vertebrates.
Class Osteichthyes (Bony fish).
Period Tertiary.
Location Europe.

 TERT

PROLATES SPP

Prolates is one of the earliest examples of the order Perciformes, which is today the largest order of vertebrates in the world. It includes such familiar fish as the sunfish, bass, darters, pike and true perches. The specimen shown is *P. herberti,* found in early Palaeocene strata in France.

Organism group Vertebrates.
Class Osteichthyes (Bony fish).
Period Tertiary.
Location Worldwide.

 TERT

OSTEICHTHYES

GOSNITICHTHYS SPP

The Green River shales of Wyoming are famous for the quantity and quality of bony fish fossils they produce. In this example, a school of small *G. paruns* have been fossilized as they died. The alignment of many of the fish indicates a possible water current, or the action of waves, before the fish were buried.

Organism group Vertebrates.
Class Osteichthyes (Bony fish).
Period Eocene.
Location USA.

LATIMERIA SPP

Believed to be extinct for 30 million years, a coelanth (the popular name) was pulled up by a trawler of South Africa in 1938. Many more have been caught since then, and all from great depths. This preserved, modern specimen shows the hallmarks of its fossil ancestors– the lobefins. *Latimeria* itself is only an offshoot of this once diverse group and it is not the direct ancestor of the first land vertebrates.

Organism group Vertebrates.
Class Osteichthyes (Bony fish).
Period Devonian–Recent.
Location South Africa.

OSTEICHTHYES

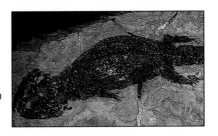

MICROMELERPTON SPP

Shortly after the amphibians evolved, they split onto two major groups. One, the labyrinthodonts, gave rise to the reptiles, while the other, the temnospondyls, remained an evolutionary dead-end. *Micromelerpeton* was a member of the latter group, and had a wide, flat head, short limbs. The species shown is *M. amphibia,* from Lower Permian deposits, in Germany.

Organism group Vertebrates.
Class Amphibia (Amphibians).
Period Permian.
Location Germany.

 PER

STENOPTERYGIUS QUADRISCISSUS

Abundant in Jurassic marine deposits, they were superbly streamlined. These fish-like dinosaurs were known as ichthyosaurs, and their teeth are commonly found. The skull tapers in a similar way to a modern dolphin's skull. Similarities between this extinct dinosaur and living, mammalian dolphins, show how two unrelated animals adapt in similar ways to similar environments.

Organism group Vertebrates.
Class Reptilia (Reptiles).
Period Jurassic.
Location Europe, North America.

 JUR

PTERODACTYLUS SPP

The first of the flying reptiles to come to light, this small species was discovered in a German quarry. The name means 'wing-finger' and the species was a late-comer to the history of this varied group. Most were relatively small, but one, named *Quetzalcoatlus* after the Aztec feathered snake god, may have had a wingspan of more than 50ft.

Organism group Vertebrates.
Class Reptilia (Reptiles).
Period Jurassic–Cretaceous.
Location Germany.

 JUR

LUFENGOSAURUS

The earliest and most primitive of the large dinosaurs, the prosauropods were ancestors of the massive sauropods. They had long tails and necks and were able to rear up on their hind legs for feeding, using their tail as a brace. The front legs, were are armed with curved claws, which would have been used against attackers. The variety shown is *Lufengosaurus*, a porsauropod of about 20ft in length.

Organism group Vertebrates.
Class Reptilia (Reptiles).
Period Triassic–Jurassic.
Location North America.

 TRI

SHUOSAURUS

The sauropods, the famous clan that included *Diplodocus* and *Apatosaurus*, were big dinosaurs, with one species that may have been more than 100ft long. But not every sauropod was a giant, *Shuosaurus* was only 30ft long and had an impressive set of teeth. The teeth were designed for cropping and chewing vegetation.

Organism group Vertebrates.
Class Reptilia (Reptiles).
Period Jurassic.
Location Asia.

 JUR

MAMENCHISAURUS

With 19 neck vertebrae, this species holds the record for the longest neck among the dinosaurs. This Asian dinosaur belonged to the group known as diplodocids, of which *Apatosaurus* (formerly *Brontosaurus*) is the most famous. It is now believed that the diplodocids and their relatives were land animals capable of rearing up, on their hind legs, to feed in the highest branches of the coniferous forests of the time.

Organism group Vertebrates.
Class Reptilia (Reptiles).
Period Jurassic.
Location Asia.

 JUR

TUOJIANGOSAURUS

The stegosaurs were a varied group of herbivorous dinosaurs in the Jurassic and Cretaceous, best known for *Stegasaurus,* with its double row of bony plates running down the back. *Tuojiangosaurus* is a recent find from the fertile deposits in China. Smaller than the 30ft *Stegasaurus,* it bore twin rows of spiky plates and a double pair of tail spines.

Organism group Vertebrates.
Class Reptilia (Reptiles).
Period Jurassic.
Location China.

 JUR

SCOLOSAURUS

Fossilized dinosaur skin is rare, providing a glimpse of their appearance and lifestyle. This fossil skin impression is from an ankylsaur known as nodosaurid – squat, barrel-bodied dinosaurs with short legs, broad backs and long tails. They were among the most heavily armored of living things, sheathed in strong plates, many having thick spikes around the body and down the tail.

Organism group Vertebrates.
Class Reptilia (Reptiles).
Period Cretaceous.
Location North America.

  CRET

TRICERATOPS SPP

It was one of the great confrontations of pre-history: the towering *Tyrannosaurus* against the heavily armoured herbivore *Triceratops*. *Triceratops'* weaponry was formidable: two curved, sharp horns angling forward from just above the eyes, with a third, shorter horn on the snout. The neck was protected by a wide head shield that flared back several feet, and was rimmed with short spikes.

Organism group Vertebrates.
Class Reptilia (Reptiles).
Period Cretaceous.
Location North America.

GASOSAURUS

The therapods began as small, meat-eating dinosaurs in the Triassic, eventually producing *Allosaurus* and *Tyrannosaurus*. *Gasosaurus* was a much smaller theropod but was built with large, powerful hind legs, tiny forelegs adapted for grasping, a long tail to counter-balance the body while running, and a mouth armed with curving knife-like teeth.

Organism group Vertebrates.
Class Reptilia (Reptiles).
Period Jurassic.
Location North America.

PROTOCERATOPS SPP

The discovery of the first dinosaur eggs, in 1923, by the Andrews expedition to central Asia, captured world attention. It has been shown that the eggs of *Protoceratops* are surprisingly common in the Gobi Desert. The eggs are usually found in circular groupings inside the remains of the hollow scrape that served as a nest. The adult *Protoceratops* was less massively built than its relatives.

Organism group Vertebrates.
Class Reptilia (Reptiles).
Period Cretaceous.
Location Central Asia.

ARCHAEOPTERYX

First there was a feather– as starling imprint found in Germany in 1860, from Jurassic rock where dinosaurs, not birds, were expected. Less than a year later, the same limestone had given up a nearly complete *Archaeopteryx* fossil. The dinosaur bones could be clearly seen. The fossil illustrated is the famous 'Berlin Specimen', unearthed in 1877.

Organism group Vertebrates.
Class Aves (Birds).
Period Jurassic.
Location Germany.

MAMMUTHUS SPP

The proboscidians, including the elephant, arose in the late Eocene or early Oligocene. The photo of a jaw of North America's *M. columbi* from South Dakota, shows the size of these extinct animals. Wooly mammoths (*M. primigenius*) from Siberia and Alaska have provided some unique fossils, with frozen specimens holding mouthfuls of grass and wildflowers.

Organism group Vertebrates.
Class Mammalia (Mammals).
Period Oligocene–Pleistocene.
Location Africa, South America.

SMILODON SPP

Known as 'Sabre-toothed tigers', these large, unusual cats were common carnivores in North and South America during the Pleistocene. The teeth look deadly and their purpose may have been either to kill prey or play a social role. The same style of dentition appeared in marsupials during the Pliocene but died out when placental mammals overwhelmed the marsupial fauna.

Organism group Vertebrates.
Class Mammalia (Mammals).
Period Pleistocene.
Location North and South America.

Index

I N D E X